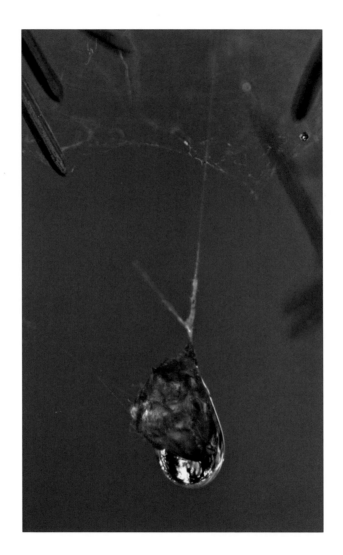

VOYAGEUR SKIES

VOYAGEUR SKIES

WEATHER AND THE WILDERNESS
IN MINNESOTA'S NATIONAL PARK

PHOTOGRAPHY BY DON BRENEMAN

WEATHER COMMENTARY BY MARK SEELEY

AFTON PRESS

THE PUBLICATION OF

VOYAGEUR SKIES
WEATHER AND THE WILDERNESS
IN MINNESOTA'S NATIONAL PARK

*has been made possible
by a major gift from:*

THOMSON REUTERS

with generous gifts from

3M Foundation

Barr Engineering

Boise Paper

Elmer L. and Eleanor J.
Andersen Foundation

Freshwater Society

Friends of Voyageurs National Park

Great River Energy

Knudsen Family Community Fund

Gordon Shepard Fund of
the Saint Paul Foundation

Martin and Esther Kellogg

Malcolm and Patricia McDonald

MN Parks and Trails Council

Jonathan H. and Martha R. Morgan

Chad Nicholson

Park Midway Bank

University of Minnesota Extension

University of Minnesota
Water Resources Center

Cover and interior design by Mary Susan Oleson
Copyediting and production assistance by Beth Williams
Printed by Pettit Network Inc., Afton, Minnesota

Library of Congress Cataloging-in-Publication Data

Breneman, Don.
Voyageur skies:weather and the wilderness in Minnesota's national park /
by Don Breneman and Mark Seeley.—1st ed.
    p. cm.
ISBN 978-1-890434-82-3 (hardcover : alk. paper)
1. Voyageurs National Park--Climate. 2. Weather—Minnesota—Voyageurs National Park. 3. Animal ecology—Minnesota.
4. Natural history—Minnesota—Voyageurs National Park.
I. Seeley, Mark W., 1947- II. Title.
QC984.M62V693 2009
508.776'77--dc22

2010027210

Printed in China

Patricia Condon McDonald
PUBLISHER

AFTON HISTORICAL SOCIETY PRESS
P.O. Box 100, Afton, MN 55001
651-436-8443
aftonpress@aftonpress.com
www.aftonpress.com

CONTENTS

FOREWORD

VOYAGEURS NATIONAL PARK—Minnesota's only national park and the only water-based park in America's precious treasury of national parks—is a wondrous place.

And Don Breneman's and Mark Seeley's **VOYAGEUR SKIES: Weather and the Wilderness in Minnesota's National Park** does a wondrous job of evoking the magic of that special mix of large and small lakes, forests and granite along our border with Ontario.

The park's 134,265 acres of forested land and 83,789 acres of sparkling waters are a place for visitors to experience nature in every season.

I have been to Voyageurs scores of times over the past three decades. I have fished there, camped there and snowmobiled there. And, of course, I have

OPPOSITE: Looking east across Rainy Lake reveals a few of the 500 islands in Voyageurs National Park.

boated there. To experience the full grandeur of Voyageurs' major lakes—Rainy, Kabetogama, and Namakan—you really must visit the park by boat.

VOYAGEUR SKIES' beautiful photos and compelling text take me back to important moments from many of my past visits to the park, and they inspire me to plan another visit soon. If you go to Voyageurs several times a year, as I sometimes have done, or if you have never visited the park, VOYAGEUR SKIES will leave you wanting to go there.

More than that, the book is intended to be the centerpiece of educational programs to be offered jointly by the University of Minnesota Water Resources Center and University Extension. It will make an admirable contribution to the education of people, young and old.

Don Breneman's text, based on a lifetime of visits to Voyageurs, tells the story of the people who have used the park's land and waters for 9,000 years: Dakota, Ojibwe, and other Native Americans; the French and British fur traders who traversed the area in the 1700s; the gold miners who came to Rainy Lake during a short-lived gold rush in the late 1800s; the loggers and commercial fishermen; and the power company engineers who built dams that still affect the flow of water through the lakes.

Tourism was the last major industry on the lakes before Congress established the national park in 1975. The restored Kettle Falls Hotel, built between 1910 and 1913, celebrates and still serves that tourism industry.

Designation of Voyageurs as a national park resulted from tireless work by former Minnesota Governor Elmer L. Andersen and others to preserve the lakes and forests of the area as part of the 3,000-mile fur traders' route from Montreal to Lake Athabasca. The late Governor Andersen was a friend and mentor to me, and I know he counted the park as a great achievement for Minnesota.

Photographs by Don Breneman beautifully capture the flora of the park: marsh marigolds, paper birch trees, the crimson bunchberry, and Minnesota's state flower, the showy lady's slipper. If you value excellent wildlife photography, as I do, you will appreciate his photos of the park's wild creatures: moose, white-tailed deer, otters, ravens, and loons.

Mark Seeley, a University of Minnesota meteorologist and climatologist, offers a season-by-season description of the effects of weather and climate in shaping the park's landscape. He also describes four major trends in climate for Minnesota and Voyageurs National

Park over the last three decades: warmer winters; higher daily minimum temperatures; increased summer humidity; and greater variation in the annual rainfall from thunderstorms.

At the Freshwater Society, we work to educate and inspire people to value, conserve, and protect water resources. **VOYAGEUR SKIES**, in its content and its goal of being an educational resource, admirably fulfills a similar mission. I applaud the book's text and photography, and I strongly recommend it to readers of all ages.

—Gene Merriam
President
Freshwater Society

• •

PRAISE for VOYAGEUR SKIES

DON BRENEMAN and MARK SEELEY capture in words and photos the special place that is Voyageurs National Park. They have woven together meteorological science with art to reveal the manner in which weather and landscapes are so intricately intertwined. The beauty of each season in Voyageurs National Park comes alive in these pages.

—Faye Sleeper
University of Minnesota
Co-Director, Water Resources Center

A white water lily floats on the calm water of Ek Lake.

BELOW: This tiny lichen, British Soldiers, gets its name from its resemblance to the uniforms of colonial British soldiers.

INTRODUCTION

AS A YOUNG BOY, traveling the lakes of Voyageurs National Park in the days of leaky wood boats and 5-horsepower motors, I developed a keen awareness of the weather and its effect on one's activities. Even today with larger, safer boats and bigger motors, the weather often is a deciding factor in daily activities in the park. Storms can come up seemingly without warning, and wind and lightning can be deadly. Extreme cold is a hallmark of the winter climate in Voyageurs National Park and the wind chill factor is serious. This is a book about the weather and climate of Voyageurs National Park and their effect on the landscape and wildlife in the park. The book highlights not only the spectacular storms and extremes but also the more subtle nuances of the weather. Mark and I hope this book will increase your awareness and appreciation of the world around you—that it will help you to see the wonder of dewdrops on a spider's web as well as the magnificence of a thunderhead lit by the glow of the late evening sun.

The landscape flora and fauna photographs included in this book take you through the seasons in this wonderful place that the glaciers swept bare millions of years ago, evolving through the ages of ever-changing climate to what we see today.

I have had a lifelong connection with Voyageurs National Park. I grew up in Littlefork, Minnesota, a small town about 30 miles west of the park. Because my parents were avid outdoors people who loved to fish, pick berries, picnic, and hunt, I was on the lakes before I was able to see over the side of a boat!

The early years were bountiful. My grandfather moved to the area in 1913 to log. He had several logging camps in the area, including one on Kabetogama Lake peninsula. From that camp, he cut and sold the logs that built Arrowhead Lodge on the Kabetogama. My father remembers fishing Kabetogama in 1917. Standing on the rocky shore, he caught northern pike with almost every cast of a bucktail spinner.

One of my earliest weather related recollections was a trip our family made to Kettle Falls when I was about four years old. We rented an 18-foot wooden boat at Ash River. With a 5-horsepower Johnson motor, we set off to visit a relative who was the dam keeper at Kettle Falls. On the way, we got caught in a violent thunderstorm and sought refuge with Mr. and Mrs. Torry on Kubel Island. Emil Torry, a commercial fisherman on

Fishing on Daley Brook with my father, Floyd Breneman, in 1948.

Namakan Lake, and his wife lived on the east end of the island. My uncle knew the Torrys well and they gave us a warm welcome. I still remember marveling at an old pump organ in their house. After the storm passed, we moved on. Mr. Torry died a few years later, but Mrs. Torry continued to live on the island into the 1970s.

Weather could be fierce in the fall. Because my dad had a partial interest in a duck-hunting cabin at the mouth of Daley Brook on Kabetogama Lake, I spent

Mrs. Torry in 1976.

many mornings shivering in a duck blind! In 1940, he and several other men were trapped during the great Armistice Day storm. When the storm swept in that afternoon, they were out hunting. Luckily, they made it back to the cabin for the night, but by next morning the bay in front of them and the brook were frozen over. The next day they walked out to the Ash River Trail. It took another day before the storm calmed and the plows were able to help them get home to Littlefork.

When you love the lake, weather is no hindrance to projects! In 1950, Mom, Dad, and I built a cabin near Gappa's Landing on Kabetogama Lake. There was no road to the cabin so we hauled all the building material in by boat. In the winter, we skied across the lake to work on finishing the interior of the cabin. After supper, we skied back to Gappa's Landing in the dark. In the dark night, you could hear the ice shifting and booming.

The ice booming on the major lakes within Voyageurs National Park indicates the stress on the ice due to the significant drop of the water level in the winter. Dams built by the paper companies at International Falls and Kettle Falls control the Namakan Lake and Rainy Lake basins. The dams raise lake levels in the summer and lower them slowly in the winter to ensure a flow of water through the hydroelectric plant in International Falls. The dam at Kettle Falls, completed in 1914, raised the level of the lakes in the Namakan basin about 3 feet. As the water level drops in the winter, the ice settles and pressure ridges develop across the lakes. These can be dangerous to cross with a snowmobile because the ice is heaved up above the supporting water. Other cracks develop along shorelines and water seeps up over the ice but under the snow. Covered by heavy snow, the water is insulated and does not freeze. These slush pockets are hazardous to snowmobilers and skiers. In spring before the runoff raises the lake levels, the lakeshore resembles the ocean at low tide.

Tornadoes are rare in northern Minnesota but they do occur. One Sunday morning, shortly after we had built the cabin, Dad and I had taken a boat to a resort to get a Sunday paper. The wind picked up as we started back toward the cabin. Looking up, we saw a funnel cloud pass directly over us! It hit a point of land on the north shore of the lake blowing over many large white pines. Another time, shortly after the park was established, a heavy wind or tornado hit Knox Island and destroyed many of its large pines.

Growing up in this natural paradise inspired my interest in photography. This led to a career of practicing

and teaching photography at the University of Minnesota. I spent as much time as I could at the cabin in the summers—more and more the fishing rod gave way to the camera. Reviewing my pictures, I realized how many of them had a weather-related theme. My colleague and weather expert Dr. Mark Seeley joins me in coauthoring this book on the climate and weather of Voyageurs National Park. We hope you enjoy the beauty and find it both inspirational and educational. Welcome to Voyageurs!

At our cabin with an early model snowmobile.

HISTORY

VOYAGEURS NATIONAL PARK comprises 218,054 acres of spectacular scenery along the Minnesota-Canadian border. It is at the western end of the chain of lakes that stretches along the Canadian border from Lake Superior to the headwaters of the Rainy River at Ranier, Minnesota. It was the water highway used by the early fur traders. The rocky shoreline is part of the Canadian Shield, some of the oldest exposed rocks in the world.

Native Americans have occupied the border region for about 9,000 years. The earliest inhabitants that left a visible record were Native Americans of the Archaic culture. The Laurel people who came after were mound builders; several mounds at the mouth of the Big Fork River bear witness to their existence. Later, about 1660, when white exploration of the region was beginning, Native Americans of the Siouan stock inhabited the area. By 1750,

Voyageurs National Park

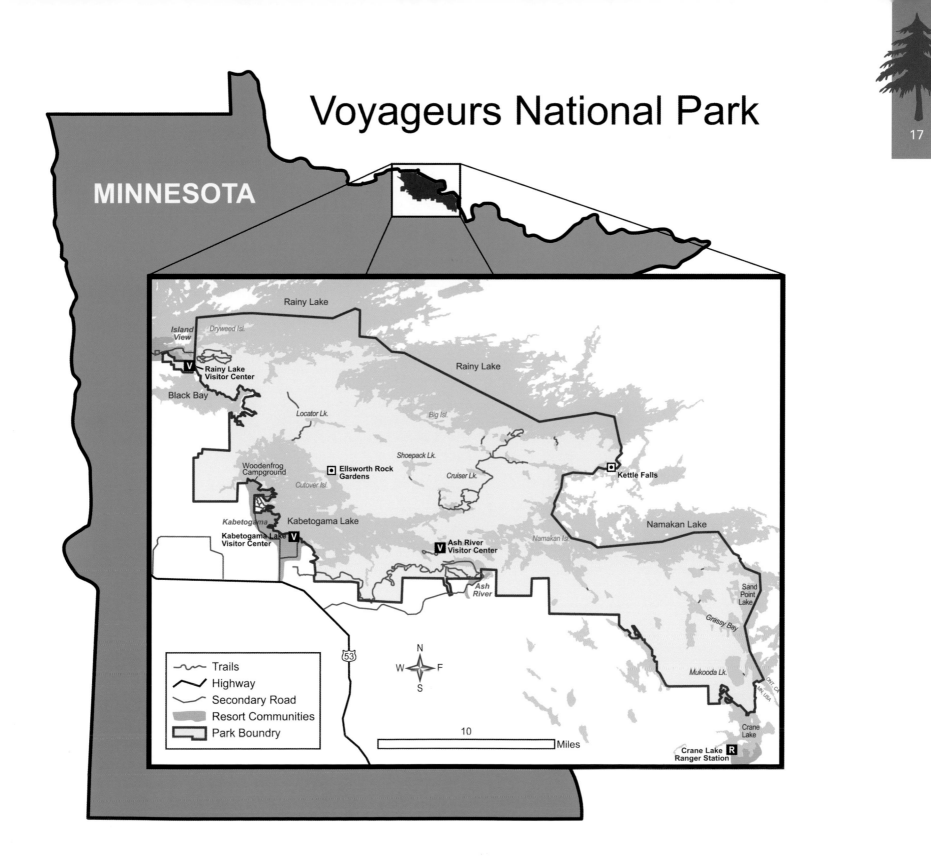

MINNESOTA

Rainy Lake

Island View

Dryweed Isl.

Rainy Lake Visitor Center

Black Bay

Rainy Lake

Locator Lk.

Big Isl.

Shoepack Lk.

Woodenfrog Campground

Ellsworth Rock Gardens

Cruiser Lk.

Kettle Falls

Cutover Isl.

Kabetogama

Kabetogama Lake

Kabetogama Lake Visitor Center

Ash River Visitor Center

Namakan Lake

Namakan Isl.

Ash River

Sand Point Lake

Grassy Bay

53

Mukooda Lk.

ONT. CA
MN USA

N
W   E
S

Crane Lake

10   Miles

Trails
Highway
Secondary Road
Resort Communities
Park Boundry

Crane Lake Ranger Station R

The rocky shores in Voyageurs are part of a geological formation known as the Canadian Shield. They are some of the oldest exposed rock in the world.

Native American people of the Laurel culture built Grand Mound, located west of the park where the Big Fork River enters the Rainy River.

the Ojibwe and other eastern tribes had forced the Sioux out onto the prairie. Today there are few signs of Native Americans in the park except for some burial sites and a rock formation on the Canadian side of Namakan Narrows. Native Americans worshipped this serpent-like vein of white feldspar as their water Manitou (god). Above the vein, painted with white pigment, is a white circular pictograph. No other white pictographs have been found along the border.

Many place names in the park reflect the Ojibwe language—Kabetogama, Namakan, and Nashata Point. Woodenfrog Campground is named for John Woodenfrog, an Ojibwe chief whose band summered on Kabetogama into the 1920s.

The 1713 Treaty of Utrecht turned ownership of Hudson's Bay over to the British, and displaced French traders crossed the Canadian borderland into what is now northern Minnesota. In 1731, the French explorer Pierre Gaultier de Varennes, sieur de la Vérendrye, established a fort at the headwaters of the Rainy River. Thus began the fur trade and the era of the colorful voyageurs (fur traders) who transported furs and trade goods from western Canada to Lake Superior by canoe. The earliest trading companies were French,

Gold miners used Gold Portage Creek, running from Kabetogama Lake to Black Bay as a short cut to the gold mines on Rainy Lake.

BELOW: Nashata Point on Kabetogama Lake is one of the sites that has a Native American name.

followed later by British and American. Various forts and trading posts existed at the headwaters of the Rainy River until about 1900.

In 1893, gold was discovered near the mouth of Black Bay on Rainy Lake. Hundreds of gold seekers flooded the area and established the town of Rainy Lake City on the east side of Black Bay. Gold Portage Creek, leading from the western end of Kabetogama Lake to Black Bay, was a shorter and safer route to the gold fields, avoiding the big waters of Rainy Lake. The largest gold mine was on Little American Island. An interpretive trail built by the National Park Service illustrates the process of mining. A second visible mine on Bushy Head Island bores directly into the side of the island just above the water line. Unfortunately, the value of the gold was not worth the cost of keeping water pumped out of the mines and by 1900, Rainy Lake City was a ghost town.

Logging was the next enterprise in the area. Logging in what is now Voyageurs National Park reached its peak in the 1920s. The Virginia & Rainy Lake Lumber Company was the major operator in the region. From its headquarters in Cusson, Minnesota, the company built logging rail lines into Namakan Lake via Ash River. Over the years, they had numerous logging camps on the lakes. Remnants of their log hoist on Hoist Bay on Namakan

The mineshaft on Little American Island, now flooded. It can be viewed from an interpretive trail on the island.

BELOW: This winch on Little American Island is a relic of the 1890 Gold Rush on Black Bay.

Lake are still visible. The Ash River Trail itself is built on an old logging railroad grade. Large iron rings driven into some rocky points on Kabetogama Lake offer evidence of the logging trade. The rings anchored log booms until flat-bottomed boats, known as alligators, could pull the booms to hoist sites. In 1929, having logged most of the valuable timber, the Virginia & Rainy Lake Lumber Company shut down operations, ending twenty years of pine logging in the area.

West of Rainy Lake, the E. W. Backus International Lumber Company headquartered in International Falls. Backus built the dams and hydroelectric plant at International Falls, along with the dams at Kettle and Squirrel Falls. The dams raised the water level of the Namakan Basin about 3.5 feet, which ensured a steady flow of water through the power plant at International Falls during all seasons. Backus also wanted to dam Lac LaCroix and other lakes further east, but the pioneer

Pilings in Hoist Bay from the Virginia & Rainy Lake logging operations in the 1920s extend far out into Namakan Lake.

conservationist Ernest Oberholtzer led efforts to stop Backus. The International Lumber Company later became the Minnesota and Ontario Paper Company and logged on the Kabetogama Peninsula until the late 1960s.

Commercial fishing began in the late 1800s for sturgeon and later for other species. The number of licenses was limited. Noteworthy commercial fishing families included the Bowmans, for whom Bowman's Bay is named and Emil Torry, who is mentioned in the introduction to this book. Fishermen on Namakan Lake sold their fish to buyers at Kettle Falls and the fish were hauled across Rainy Lake to the town of Rainer and shipped east by rail. Commercial fishing ended on Kabetogama Lake in 1923 and on Namakan and Rainy Lakes in the 1960s.

Prohibition added color to the area. Blind Pig Channel on the eastern end of Namakan Lake named the place

The railroad to Hoist Bay crossed Moose River on these timber pilings.

Iron rings driven into rock points were used to anchor log booms. This one is at the mouth of Mud Bay, Kabetogama Lake.

BELOW: Walleye fishing has long been a major tourist attraction for the region.

where American bootleggers met Canadian counterparts to smuggle liquor into the United States. A "blind pig" was a slang term for illegal liquor operations. Another blind pig was on Little Martin's Island on Kabetogama Lake; no doubt there were many more. Jug Island on Kabetogama acquired its name from commercial fishermen hiding their bottles on the island.

Tourism was the last industry on the lakes before the establishment of Voyageurs National Park. Every major lake and some smaller ones with road access hosted resorts. Walleye fishing was the main attraction and still draws many people to the park. Tourists from Illinois and Iowa were early visitors to the area and still comprise significant numbers of guests at the remaining resorts.

Two additional attractions catered to tourists. On Kabetogama Lake, Jack Ellsworth, a contractor from the Chicago area, built a large rock garden. Having built a retirement cabin on the north shore of Kabetogama Lake, he cleared off a large rock hill, terraced gardens, and planted hundreds of tiger lilies. Later he built rock sculptures out of local stones. Scores of visitors stopped at the gardens each summer. The National Park Service has preserved the site and even without as many beautiful flowers, it still is a popular stopping point.

Ellsworth Gardens had over 13,000 lilies blooming during its hey-day in the early 1960s.

BELOW: The view from the top of Ells-worth Rock Gardens highlights the rock sculpture and the well-kept grounds.

Kettle Falls Hotel was built in the early 1900s and still accommodates visitors. The National Park Service renovated it in 1987.

A second historical attraction is the Kettle Falls Hotel. Built between 1910 and 1913 as a stopover point for travelers going between International Falls and Crane Lake, this two-story wood building has a long and colorful past. In the logging days, it was part brothel, part bar, and part restaurant that separated many lumberjacks from their pay! Renovated by the National Park Service in 1987, the hotel is still a popular destination for dining and lodging.

In order to expand tourism and to preserve the Kabetogama Peninsula, the U.S. Congress established Voyageurs National Park in 1975. Former Minnesota Governor Elmer L. Andersen led the preservation effort. Voyageurs National Park preserves a segment of a 3,000 mile waterway from Montreal to Lake Athabasca. Unique in that it is the only national park primarily accessible by water, Voyageurs provides visitors miles of spectacular scenery, a colorful past, and rich opportunities to observe nature in a nearly pristine state.

Read on—COME AND VISIT IT WITH US!

Two canoeists enjoy the quiet solitude of Ek Lake. The park service provides canoes on inland lakes to avoid possible contamination from invasive species.

WEATHER

VOYAGEURS NATIONAL PARK lies within Koochiching and St. Louis Counties along a 55-mile stretch of border between Minnesota and Ontario, Canada. It encompasses 134,265 acres of land and 83,789 acres of water, for a total of 218,054 acres. The landscape is heavily forested in many places with slightly variable relief. Jack pine, aspen, and oak dominate the landscape. The dominant soil association is Newfound-Mesaba-Quetico soils formed in thin noncalcareous loamy glacial drift overlying bedrock. Within its boundaries there are four well-known lakes: Rainy, Kabetogama, Namakan, and Sand Point along with numerous other smaller lakes and streams. To a degree the soils, vegetation, and lakes influence the climate of the region. Because of the sparse population density and limited number of year-round residents the available climate stations from which to derive a reasonable description of the environment is very limited.

Records from only nine climate stations in or nearby Voyageurs National Park are the only source of historical data to quantify the environment of the park. These include observers at Crane Lake (1926–77), Orr (1926–72), International Falls (1897–2009), Kettle Falls (1943–2009), Big Falls (1930–2009), Little Fork (1909–2009), Winton (1913–95), Ely (1911–2009), and Kabetogama (1997–2009). The climate records from these locations vary in length, type of measurement, and continuity. Such a sparse quantity of data to cover a landscape of 340 square miles limits the rigor of doing very precise spatial and temporal analysis of patterns in temperature and precipitation. Nevertheless a general description of climate characteristics, their seasonal dimensions, and a look at extreme weather events and climate episodes can be done.

PRECIPITATION (SNOWFALL AND RAINFALL)

Mean annual precipitation across Voyageurs National Park ranges from 26 to 29 inches. This is considerably greater than the prairie landscape of western Minnesota which receives 19 to 23 inches, but it is less moisture than those areas to the east along the north shore of Lake Superior where annual precipitation ranges from 30 to 32 inches. Maximum annual historical amounts of precipitation have topped 38 inches, but only in a couple of years. Rainfall deficiency produced significant drought

in 1910, 1929, 1931, 1933, 1936, 1940, 1952, 1956, 1958, 1976, 2003, and 2006. In some of these years forest fires burned in and around Voyageurs National Park, especially when the total annual precipitation was less than 18 inches.

The precipitation pattern across the park is affected by prevailing wind direction, minor elevation differences, and lakes within the area. On the lee edge of Voyageurs National Park at Crane Lake precipitation is undoubtedly enhanced by the water vapor released upwind from the three major lakes. This is evident in the higher precipitation values found there. Roughly two-thirds of the annual precipitation falls during the months of May through September much of this in the form of thunderstorms. Over the months of June through September, average monthly values exceed 3 inches, but have been as high as 11 to 12 inches in individual years. June is usually the wettest month of the year in terms of both quantity (3.75 to 4.25 inches) and frequency of precipitation (typically 12 to 14 June days bring measurable rainfall). It is not uncommon to get thunderstorm rainfalls of greater than 3 inches, and on rare occasions amounts over 4 inches have been measured. Some of the wettest years in history were 1941, 1944, 1968, 1977, 2001, and 2005. February is the driest month of the year (0.60 to 0.70 inches), but usually brings several significant snowfalls.

The average annual days when measurable precipitation occurs is 130 suggesting a precipitation frequency of about one day in every three. In the very wet year of 1977 precipitation was reported on 160 days, nearly 44 percent of the time.

Seasonal snowfall deposition is highly affected by which way the wind blows and whether or not the area lakes are ice covered. On the lee side of the lakes, Crane Lake reports higher average snowfall values in November (average 10 inches) and December (average 15.2 inches) due to enhanced water vapor transport from the upwind unfrozen lakes. Precipitation deposition rates on the landscape are also higher in the uplands, which range in elevation up to 1,500 feet (compared to some areas that are only 1,200 feet). Mean seasonal snowfall ranges from 50 to 70 inches within the park boundaries, but some locations have recorded in excess of 100 inches on a number of occasions. International Falls recorded over 100 inches in the snow seasons of 1988–89 (106.5 inches), 1991–92 (111 inches), 1995–96 (116 inches), and 2008–09 (125.7 inches), while Crane Lake recorded 103.8 inches in 1969–70, and Orr recorded 109.2 inches in 1965–66. Little Fork reported 103 inches of snowfall in the 1995–96 season as well. Relative to the rest of Minnesota, Voyageurs National Park occasionally sees very early autumn snowfall and very late spring snowfall.

Earliest autumn snowfall occurred on September 14, 1964, when 0.1 to 0.3 inches was reported. October snowstorms have occasionally brought over 8 inches, as was the case on October 29, 1932, when Orr reported 8.5 inches. Conversely, significant spring snow has been reported as late as May 28, when in 1947 the observer at Orr reported 4 inches.

Voyageurs National Park often reports the longest lasting and deepest snow cover in the state of Minnesota. The average duration of at least 1-inch snow cover ranges from 150 to 160 days across the park. It has been as long as 169 days as it was in the winters of 1955–56 and 1995–96. Snow-bearing storms on average start in late October, but can occur as early as the second half of September. At least a trace of snow has been recorded in every month but July and August. Maximum snow depth usually occurs in the month of March and often exceeds 30 inches. Snow depths exceeded 40 inches during the winters of 1965–66 and 1968–69, making it difficult for deer to browse for food. Late fall and early winter snowfalls are enhanced by the open waters of Kabetogama, Namakan, and Rainy Lakes, but once the lakes become ice covered they have little effect on snowfall rates. The date for the loss of winter snow cover has ranged from the first week of April to the first week of May over the years, depending on the severity of the winter and early spring temperature conditions. On occasion lakes may not be ice free for the fishing opener the second weekend of May.

During the warm season, thunderstorm trajectories across Voyageurs National Park are most frequently northwest to southeast or southwest to northeast. The area records 30 to 35 days with thunderstorms each year and some are very intense. Daily rainfall amounts occasionally exceed 4 inches from these intense thunderstorms, but more commonly they are moving along so fast rainfall amounts typically range from 0.5 to 1 inch. Thunderstorms sometimes bring hail to the park, but only about once every three years. The severe weather season rarely brings tornadoes, but can often bring strong thunderstorm winds of 60 mph or greater. Tornadoes were reported near Voyageurs National Park on August 6, 1969, but none were sighted within the park. In addition, strong winds were observed on July 4, 1999, associated with the straight-line wind storm that devastated the Boundary Waters Canoe Area, but little wind damage if any was reported from Voyageurs National Park.

TEMPERATURE

Because of the complex ecosystems and abundance of surface waters in Voyageurs National Park the spatial and temporal variability of temperature is as great here as anywhere in the state due to all of the microclimatic effects. Southern Minnesota locations are much warmer in all seasons of the year, while western Minnesota counties are warmer in the summer, but just as cold in the winter months.

Moderated by all of the surface water, the mean annual temperature ranges from 36 degrees F to 38 degrees F within the park. These are some of the lowest annual values in the state. Winter brings a dominance of a northwesterly wind pattern coupled with the presence of snow cover, which can produce some remarkably cold and dangerous weather conditions. Wind chill readings in Voyageurs National Park have been as cold as the -60s F, where exposed skin can freeze in just a few minutes. Layered clothing, hooded parkas, gloves, and insulated boots are the normal attire in winter for year-round residents, especially those who do ice fishing or snowmobiling on the frozen lakes.

Winter winds from the north usher in polar high-pressure systems, but they rarely linger for more than a week. The longest stretch of below 0 degrees F readings was probably in 1925 when the temperature remained below 0 degrees F for 18 consecutive days during December. As recently as 1994 and 1995 park temperatures remained below 0 degrees F for a week or longer, but in the past two decades this has been rare. Most of the extreme cold readings have taken place in the past. In January of 1909 temperatures fell below -50 degrees F for three days, bottoming out at -55 degrees F on January 6 when the daytime high only reached -29 degrees F. Temperatures even colder than this have probably occurred within the park but in low-lying sheltered areas away from lakes.

Freezing temperatures have occurred every month of the year. In fact, among statewide mid-summer temperature extremes Kabetogama reported a low of just 23 degrees F on July 8, 2000, a statewide record value for that date. The length of the freeze-free growing season has ranged from less than 60 days to 140 days, with a median value of 100 to 110 days. Standard deviations of daily temperature are highest in the winter months, ranging as high as 14 to 15 degrees F. They are lowest in summer, at 6 to 7 degrees F.

The longer nights of winter and relatively modest wind speeds mean that the effects of landscape condition (dry, wet, snow covered) and position have a huge effect on temperature, and combine with dramatic changes in air mass to produce the higher variability in the winter months. Late winter and early spring have sometimes brought a temperature change of over 60 degrees F in less than 24 hours. For example as recently as January 31, 2009, the afternoon high temperature at Kabetogama was 41 degrees F, but by the next morning, February 1, 2009, the temperature had plummeted to -27 degrees F. Obviously in the summer, longer days and less frequent air mass changes lead to smaller daily temperature variability. This is also the case during extremely cloudy periods when air mass stagnation and abundant surface moisture from the lakes keep a low cloud deck in place for periods of days. Over March 26 and 27, 1925, the temperature only varied by 1 degree F, from 41 to 42 degrees F over a period greater than 48 hours.

Mean diurnal temperature ranges within Voyageurs National Park are commonly from 20 to 25 degrees F for most months. At some relatively exposed inland locations diurnal temperature range approaches 30 degrees F. Peak values occur consistently in May, when lingering lake ice and snow cover can hold down overnight minimums, while a rapidly increasing sun angle and warm air advection from southerly winds can produce unusually warm days as well. Relatively low diurnal temperature ranges occur in November as a result of the increased wind speeds and a much higher incidence of cloudiness, both of which moderate temperature changes.

On average the summer season brings four to five days with daytime highs that reach 90 degrees F or greater. There have been occasional heat waves such as 1955 and 1988 when the summer brought 22 and 26 days with 90 degrees F or greater. In recent decades unusually high summer dew points have been recorded at Voyageurs National Park. In July of 1995, 1999, and 2001 dew points in the 70s F elevated the daily Heat Index Value (combined effect of temperature and dew point condition expressed as human comfort) to make the outside air feel like it was 100 degrees F. An actual air temperature reading of 100 degrees F is pretty rare in Voyageurs National Park, though it has been as hot as 106 degrees F on July 12, 1936. The last time it reached 100 degrees F there was on July 19, 1977.

The late evening sun highlights orange lichens growing on the northwest point of Chase Island, Kabetogama Lake.

Early in April, the lake ice begins to thaw along shorelines and in shallow bays. Because the water level in the major lakes is lower during winter, the early spring shoreline resembles an ocean at low tide.

Thawing ice crystallizes or honeycombs and loses its structural integrity. Even though the ice still may be a foot thick, it shatters easily when it strikes a hard object.

A fishing spider walks over the disintegrating ice crystals as life awakens for a new season in Voyageurs.

BELOW: A hole in the thawing, spring ice reveals a large boulder on the lake bottom. Free of algae that will come later, the water is exceptionally clear.

OPPOSITE PAGE: Wind shelves the disintegrating ice pack up on shorelines. A high wind can push the ice up several feet and speeds the disintegrating ice-out process.

Mountain maple blossoms are a brilliant red in early spring; however, if you look closely at a mountain maple in any season, you usually can find something red on the tree.

OPPOSITE PAGE: Maple blossoms and open water announce spring in Voyageurs!

A white-tailed doe and fawn enjoy the moderating spring temperatures and freedom from the heavy snow cover of winter.

A young moose pauses along the shore of Daley Brook before resuming browsing on the budding willow and alder brush in the background.

Aspen trees reach for the warm spring sunshine. Aspen trees leaf out in early May and form a distinct canopy by Memorial Day.

River otter patrol the calm spring water of Kabetogama Lake. Voyageurs is home to otter, mink, fisher, pine martin and ermine—all members of the weasel family.

BELOW: A hungry mink enjoys a crawfish "shore lunch." Crawfish is one of the primary food sources for mink.

Bunchberry is a variety of herbaceous dogwood that grows throughout the park. Late in the summer the plant will have striking red berries.

Blueberries bloom in late May and early June. Because the blossoms are susceptible to frost, a hard spring frost can rob bears of one of their primary summertime food sources.

Spring run-off swells Clyde Creek as it cascades toward Kabetogama Lake.

Marsh marigolds bloom in profusion in wet swampy areas throughout the park. They usually bloom in late May but, depending on the weather, the date for blooming can vary as much as two weeks.

Brown bark peels from a young paper birch revealing the characteristic whiter bark.

This mother loon has a chick by her side and an egg resting in the nest in the background.

A thunderhead looms on the horizon. Late May often brings several severe thunderstorms, some even accompanied by hail.

Raindrops decorate the tips of balsam fur needles after a May rain.

The star flower grows in wet shaded woods. It is rare in the plant world for a blossom to have seven petals.

BELOW: Tiny mushrooms reach up through the moss on the forest floor for their moment in the sun.

Rain and hail are common in late May; occasionally, a funnel cloud has been reported in the park.

SUMMER

METEOROLOGICAL SUMMER is June through August, the period when Voyageurs National Park sees its largest number of visitors. Virtually all the water bodies in the park are dotted with water craft of all sizes and the campgrounds are typically busy each night of the week. June days are long, with up to 16 hours and 8 minutes of daylight, and temperatures are typically in a comfortable range from the low 70s F during the day to the low 50s F at night. Weekly thunderstorms bring rainfall to the park, mostly in the late afternoon and early evening. Climatically, June is the wettest month of the year averaging about 4 inches in total, mostly from 6 to 7 thunderstorms that occur during the month. Boaters and fishermen need to remain aware of the weather during the day since thunderstorms with heavy rain, hail, and strong winds can emerge in a matter of hours. They keep their eye on the western horizon periodically, and many use a portable NOAA Weather Radio to keep abreast of any weather threats. With broadcast towers in

International Falls, Orr, and Ely, the weather forecasts and warnings from the National Weather Service can be picked up on portable NOAA Weather Radios just about anywhere in the park.

Frosts often still occur in the month of June. More than 40 percent of all years bring a frost on at least one day in June. The most recent June frosts occurred in 2004 and 2009. Though negligible in quantity, snowfall was observed in the park over June 2 and 3, 1945, and again on June 2, 1969 (0.3 inches). Far more commonly June afternoons bring temperatures in the 70s and 80s F. There is usually at least one June day that reaches 90 degrees F or higher. In 1988 there were eight such days, while in 1910 there were ten days of 90 degrees F or higher. In the shallow bays of the lakes water temperatures climb into the 70s F, suitable for swimming and water skiing by the end of June.

Extreme measures of June climate include a thunderstorm rainfall of 4.15 inches on June 24, 1898, (storm total 5.45 inches); an afternoon high of 101 degrees F on June 27, 1912; and a morning low of just 21 degrees F on June 1, 1964. The driest June in history brought just 0.70 inches in 1961. The wettest June occurred in 1944 when a total of 9.46 inches was reported from 14 days with precipitation.

During July, Voyageurs National Park offers vacationers a refuge from the mid-summer heat that dominates many other parts of the Minnesota landscape. With the exception of the Lake Superior shoreline, all other locations in the state report average daytime highs in the low to mid 80s F with overnight lows in the 60s F. The camping and fishing destinations within Voyageurs National Park boundaries report daytime highs in the upper 70s F and comfortable nighttime temperatures in the low to mid 50s F. Air conditioning is little used in and around the park. It can on occasion get hot in July, but it is usually a short-lived spell. A daytime high of 100 degrees F or greater has been reached only on 18 days dispersed in the summers of 1912, 1923, 1931, 1932, 1933, 1936, 1975, 1976, and 1995. Most of those readings have been in July, with the most notable heat wave from July 9 to 16, 1936. During that eight-day spell the average daily high was 100 degrees F, while the average overnight low was a more tolerable 65 degrees F thanks to the cooling effect of the lakes. In fact, although the overnight lows elsewhere in Minnesota during July can sometimes be as high as the upper 70s to low 80s F, the cooler lake water within the boundaries of the park tends to keep even the mid-July nighttime temperatures in the more comfortable 50s and 60s F. Over a century of climate records for the area indicate that the overnight low has stayed warmer than

70 degrees F during July less than 25 times (less than 1 percent of all days).

Unlike June and August, frosts are almost unheard of during July. The only historical evidence in the climate records are readings of 32 degrees F on July 11, 1911, and on July 2, 1972. About every other year July overnight temperatures dip into the 30s F, but do not lead to frost.

Though more often a drier month than June, July typically brings 8 to 9 thunderstorms across the park. Some of these can deliver 3 to 4 inch rainfalls in a single day, along with wind gusts of over 60 mph. In addition about every other July brings a hail storm. On occasion July brings rainfall almost every other day to the park like it did in both 1948 and 2001, resulting in over 8 inches. All day rains are extremely rare as most storm systems develop in late afternoon or early evening and move rapidly across the area. In some years July can be exceptionally dry, as in 1941 when it only rained three times and all other days were bright and warm.

Extreme measures of July climate include a thunderstorm rainfall of 4.75 inches overnight on July 29–30, 1948; an afternoon high of 106 degrees F on July 12,

1936; and a morning low of 32 degrees F on July 11, 1911, and July 2, 1972. The driest July in history brought just 0.02 inches on a single date in 1930. The wettest July occurred in 1966 when a total of 9.52 inches was reported, mostly from intense thunderstorms on July 2 (4.20 inches) and on July 31 (2.93 inches).

August is typically the peak of the water recreation season as the lake temperatures are at their highest from accumulated summer heat. Air temperatures start to taper off a bit from July, as the average daily high temperatures are in the mid-70s F. Individual days can still reach the 90s F, and even strings of such days can produce a mini-heat wave of sorts. Such was the case over August 19 and 20, 1976, when both days hit 99 degrees F, as well as the first three days of August in 1989 when each day reached the 90s F. Seven August days reached the 90s F back in 1984, 1988, and 1991. Campers and boaters have to be prepared for some cool nights in August as overnight lows average in the low 50s, but often fall off into the 40s F and sometimes even the 30s F. When overnight low temperatures get this cold there is often fog formation over the lakes. August frost occurs in Voyageurs National Park about once every eight to ten years. Only 1907, 1982, and 2004 brought more than one frosty night in August.

Thunderstorms occur typically over six to seven days in August, and heavy rains are still a possibility, especially in the late afternoons and early evenings. Thunderstorms produced over 3 inches of rain on a single August day in 1942, 1944, 1988, and 2001. Though tornadoes and funnel clouds are exceptionally rare in the area of Voyageurs, these large thunderstorms are sometimes accompanied by short-lived strong downburst winds, or even straight-line winds (derechos) that can level trees, flip boats, and take the roofs off cabins. Such was the case on August 22, 1995, when afternoon thunderstorms brought hail and wind gusts up to 67 mph.

Extreme measures of August climate include a thunderstorm rainfall of 4.82 inches on August 30, 1942; an afternoon high of 99 degrees F on August 19, 1976; and a morning low of 25 degrees F on August 28, 1934. The driest August in history brought just 0.15 inches in 1930 when it only rained on three days. The wettest August occurred in a statewide drought year, 1988 (as many counties reported severe to extreme drought conditions), when a total of 12.15 inches was reported within Voyageurs National Park. In August of 1988 as 22 separate days brought measurable rainfall in what was undoubtedly a frustrating month for vacationers to the park.

Juneberries grow on small shrubs. They have much the same flavor as blueberries and are a favorite food for bear.

BELOW: A young white-tailed buck pauses on his way to the lakeshore. When the horseflies and deerflies are bad in June and July, deer often wade into the water to protect their legs and underbellies.

A great blue heron passes a warm afternoon on a rock point near Gold Portage Creek. Haziness caused by evaporated water vapor screens the far shoreline. There also is some concern that air pollution could be affecting air quality along the U.S. and Canadian border.

A common loon rests after coming up from a long dive. Loons can dive as deep as 250 feet and can remain underwater for up to 5 minutes. The loon is the Minnesota state bird. The wailing call of the loon is one of the most delightful sounds in Voyageurs.

BELOW: A doe and fawn mirror each other on a sandy beach near Daley Brook.

A bald eagle and a dragonfly share the sky in Voyageurs. Both a part of the web of life in the North Country, they are predators, but on a vastly different scale.

The showy lady's slipper blooms during late June and early July. The Minnesota state flower prefers cool, damp swampy areas. It is one of 43 orchid species in Minnesota.

Blue flag iris decorates the shoreline of Lost Bay. This flower blooms in June along the marshy edges of lakes and ponds.

A beaver enjoys a "shore lunch" near the mouth of Daley Brook. These large rodents are second only to humans in their ability to alter their environment.

A bald eagle lifts off from atop a tall white pine, a favorite perch for eagles. These large predators can have a wingspan of up to 90 inches.

A wild calla lily graces the shore of Agnes Lake in June. Agnes Lake is a short hike from Lost Bay on Kabetogama Lake.

A white-tailed buck, antlers covered in velvet, comes down to the lake for an evening drink.

This white-tailed doe is still shedding her grey winter coat in early July. You can see the reddish summer coat coming in under the old hair.

A wind shear aloft disperses dark clouds to the north. South winds are dominant in the summer months.

Thunderheads loom over Kabetogama Lake in mid July. Because surface moisture is taken up during the heat of the day, the prime time for thunderstorms is from 4:00 p.m. to 9:00 p.m.

A convective thunderstorm complex showing a distinctive out-flow boundary blocks the evening sun as it approaches from the northwest.

Evening sun illumines this orange lichen-covered rock island. Orange lichens grow predominately on the northwest-facing rock surfaces.

Borderland is famous for its sunsets. The east/west orientation of the border lakes provides great opportunities for viewing dramatic sunsets.

A white pelican takes flight in late evening. Voyageurs has nearly 17 hours of sunlight in June with light in the evening sky until after 11:00 p.m.

The evening sun is framed by willow branches on the north shore of Chase Island, Kabetogama Lake. As fall approaches, days become shorter and sun sets earlier.

Scud clouds or rogue clouds are ragged edged low-level clouds that are usually fast moving. They may be remnants of larger thunderstorm clouds that have broken up and dissipated, especially in the shortening days of late summer.

Cloud streaks can produce dramatic sunsets in late July and early August.

What mariners call a mackerel sky creates a far different looking sunset.

A pelican glides in for a landing on a warm summer evening. The number of white pelicans has increased dramatically in recent years.

AUTUMN

THE AUTUMN SEASON BRINGS an earlier and more abrupt change to Voyageurs National Park than to most other sections of Minnesota. Though warm, muggy nights and even thunderstorms can still occur, many nights turn crisp in September and there are often a few frosty mornings. On average there are three September frosts, but there have been as many as ten (1942 and 1955), with lows as cold as the teens F. As a result of differences between lake temperatures and the colder overnight air temperatures, morning fog is not uncommon. But it often burns off by mid-morning as the sun is still quite strong during this equinox month. About once every seven to eight years snows occur in September. Measurable snowfall has come as early the 14th, though a snowfall this early always melts. The last significant September snowfall was in 2003, when a couple of inches fell.

Fall leaf color change commences in late September and is accelerated by the number of nights that bring temperatures below 40 degrees F. Former state climatologist Earl Kuehnast found that peak fall color display is associated with an accumulation of seven to ten nights with temperatures in the 30s F. The mix of deciduous (birch, maple, aspen) and coniferous trees (white and red pine, spruce, and fir), along with the variety of understory plants (dogwood and sumac) make for a wonderful bouquet of colors. In addition there is a tendency for the weather to deliver more clear days, so that the colors are not dulled by cloudiness. In fact clear day frequency is near a maximum during the last days of September and beginning of October. Visitors are treated to fall colors, and to migrating birds, and watching bald eagles feed on the tulibee, lake herring, and cisco in shallow bays.

Extreme measures of September's climate include: a snowfall of 1.4 inches on September 30, 1981; an afternoon high of 99 degrees F on September 7, 1976; a morning low of 12 degrees F on September 18, 1929; and a daily rainfall of 4.78 inches on September 10, 1961. The driest September was in 1897 with just 0.20 inches, all coming on one day, the 15th. The wettest September was in 1927 with 9.92 inches, totaled from 17 rainy days.

October brings even more frequent frosts and snows. It has snowed as much as 15 inches in October. On October 22, 1995, it snowed over 5 inches, a precursor to a long and snowy winter. Other years with significant October snowfalls were 1932, 1942, and 1981. October low temperatures can fall to below 0 degrees F values. On October 26, 1936, it fell to -12 degrees F. In most years the lakes are still too warm to freeze up in October. Nevertheless thin ice may occasionally form in the bays and inlets.

Following the first strong autumnal storms, October skies are filled with migrating birds from Canada. The cloud types change from vertical cumulus cloud forms to stratus or layered clouds. In addition as air temperatures fall, the clouds form lower in the sky and may blanket the sun for longer periods. Wind speeds increase and stir the waters of Voyageurs National Park more frequently, often into white caps. The low sun angle combined with brilliant fall colors, active wildlife, and partly cloudy skies attract outdoor photographers. But sometimes the fall color season is short-lived because the stronger winds of October rip the red and yellow leaves from the deciduous trees.

Extreme measures of October's climate include: a snowfall of 8 to 9 inches on October 29, 1932; an afternoon high of 89 degrees F on October 5, 1963; a bone-chilling morning low of -12 degrees F on October 26, 1936; and a daily rainfall of 2.85 inches on October 27, 1971. The driest Octobers were in 1933 and 1937 when portions of Voyageurs received zero precipitation. The wettest October was in 1971 when 17 days brought rain and the monthly total was 9.60 inches.

By November, the landscape starts to freeze up. Most days are cloudy, sometimes even foggy and it rarely warms above 50 degrees F. Stratoform (layered) clouds dominate the area and often persist all day. Being the cloudiest month of the year, November brings an average of only three to four clear days. Average overnight lows are in the teens F so the ground begins to freeze by the third week, and ice cover forms on smaller lakes. Most bays in the big lakes (Rainy, Kabetogama, and Namakan) begin to ice up by the last week. Mobility of the wolf pack increases as they can travel more easily from island to island across the ice, hunting for food.

November is also a windy month, trailing only April in terms of average wind speed. The first winter storms of the season usually cross Voyageurs National Park and often bring winds greater than 30 mph. In extreme cases wind gusts may range between 45 and 50 mph. These storms produce significant snowfall. Most observers report 8 to 12 inches of snowfall during the month, and it has snowed as much as 15 inches in one day, as it did at Kabetogama on November 24, 2003. In November of 1955 six snow-bearing storms crossed the park bringing over 30 inches.

Extreme measures of November's climate include: a snowfall of 18.5 inches over November 9–10, 1977; an afternoon high of 75 degrees F on November 5, 1975; a low of -34 degrees F on November 30, 1964; and daily precipitation of 2.62 inches (with 6 inches of snow) on November 3, 1919. Driest Novembers (1924, 1928, 1939, 1999, and 2002) brought zero precipitation to all or parts of Voyageurs National Park. The wettest November was in 2005 when 4.87 inches was reported, including 16 rainy days.

Autumn can be lengthy, serene, and full of beauty. Aboard a boat in the late autumn season, watching the streams discharge fresh water into the lake, listening to the ancient Earth sounds of waves lapping the rocky shores, and the wind rustling the trees as they let go of their beautiful leaves evokes an unmatched feeling of tranquility and provides a lifelong memory to those who have visited Voyageurs National Park at that time of year. Many local residents refer to the autumn as their favorite season. Two of the most exceptional autumns occurred in 1916 and 1987. In those years few days brought precipitation, while most brought bright sunshine and afternoon temperatures in the 40s and 50s F. They say boats were brought ashore very late in the season in those years as most residents and visitors wanted to enjoy late autumn fishing, as well as the scenic views and nature watching.

Crimson bunchberry leaves contrast with other rotting vegetation and still-green yarrow. Early morning temperature inversions intensify the pungent odors emitted by the decaying vegetation.

Fall colors on Sphunge Island, Kabetogama Lake, create a brilliant panorama.

Common mergansers form large flocks in early fall to hunt minnows. The mergansers often are accompanied by ravenous gulls. Mergansers migrate south just before lakes freeze in November.

Maple and aspen leaves float over a small riffle on Clyde Creek, Kabetogama Lake.

Cooler fall air temperatures contrast with warmer lake water and bring morning fog.

White birch are mirrored in the quiet waters of Ek Lake. Ek, one of several inland lakes on the Kabetogama Peninsula, can be reached via a short hike.

Reeds glow in the low sun on a hazy, fall afternoon.

A ruffed grouse struts on a late fall day. These elusive birds are more visible after the leaves have fallen.

LEFT: The dark fall sky spotlights a scarlet maple on the shore of Hoist Bay, Namakan Lake. Cooling temperatures cause a lower cloud ceiling in October and November.

A lone Mountain Maple contrasts with pine near the mouth of Ash River.

Damage to this mountain maple leaf reveals the stress of a long growing season. Probably the most colorful tree in the park, its leaves turn red as early as late August. Mid to late September usually is the best time to see color in the park.

Shortening days and frost paint blueberry leaves in lavender hues. Small shrubs like this often hold their color until mid to late October.

The secluded waters of Clyde Creek carry leaves down to Kabetogama Lake on a quiet fall afternoon. This small stream originates in a beaver pond about a quarter mile from the lake and its water volume varies, depending on the time of year and the amount of rainfall.

A pair of mallards test early ice on Duck Bay, Kabetogama Lake. Ice forms first in shallow bays in early November and gradually works its way out to cover the entire lake in December.

RIGHT: A tamarack is resplendent in late fall color along Daley Brook. Tamarack are the last tree to turn color in the fall.

Aspen, birch, and maple leaves end the season in a quiet pool.

BELOW: Hoar frost hints of colder weather to follow. Freezing temperatures have been recorded during every month of the year in Voyageurs.

Morning fog filters the rising sun on a cold fall morning. The relatively warm lake waters readily surrender moisture into the colder and drier overlying air, creating wispy, mystical waves of vapor visible above the surface. Photo by Joe Jovanovich.

A crescent moon hangs in the clear fall sky over Sphunge Island, Kabetogama Lake.

WINTER

METEOROLOGICAL WINTER IS MOST often defined as the December through February period, even though the snow accumulation season for Voyageurs National Park most often begins in October, sometimes even earlier. In 1912, 1951, 1964, 1974, 1981, 1992, 1993, and 2003, September brought measurable snowfalls. Snow also lingers late into the spring in most years, frequently in April, and not uncommon even in May. Measurable snowfalls in May occur in about four years out of ten. In 1954 May snowfall totaled 13.4 inches capping off a snow season that brought nearly 94 inches to the shores of Rainy Lake. The most recent May snowfall was May 16, 2009, when 0.3 inches was reported. The snow season of 1964–65 was probably the longest in history for Voyageurs National Park, with the first snowfall on September 14, 1964, of 0.3 inches, and the last on May 28, 1965, of 0.3 inches, 256 days later!

The winter recreation season, including ice fishing, snowmobiling, and cross-country skiing usually begins well before the winter solstice. More often than not, snow begins to accumulate on the ground during the first ten days of November, and in an average year, snow accumulation will continue until the first week of April. The persistence of snow cover is a desirable attribute for winter recreation, and within Voyageurs National Park the ground is covered with snow for 150 to 160 days. Though average total snowfall ranges from 60 to 75 inches, extreme winters have brought 115 inches or more (1995–96 and 2008–09). Average maximum snow depth ranges from 18 inches to over 24 inches and typically occurs in late January or early February. There have been winters when snow depth has exceeded 36 inches and contributed to high mortality rates in the deer herd population. This occurred in the winters of 1965–66, 1981–82, and 1995–96. In addition ice cover forms on the park's lakes typically by the third week of November and is often safe enough for ice fishing by early December when a network of ice roads begins to appear on the lakes. The ice roads are in common use typically until March.

Because abundant snow cover reflects a significant amount of the sun's energy, little is absorbed to heat the landscape. Compounded by the short days and low sun angles, there is a shortage of radiant energy that can be converted to heat during the winter months. Near the winter solstice, there is less than 8 hours and 20 minutes of daylight. As a result, many daytime high temperatures remain below 0 degrees F (commonly 10 to 12 days each winter), while nighttime lows fall well below 0 degrees F quite often (45 to 50 nights on average). In such cold temperatures with little wind, the air contains suspended ice crystals that produce interesting optical effects, including sun dogs, sun pillars, and beautiful sunrises and sunsets.

Winter temperatures in Voyageurs National Park are legendary and among the coldest in the state. Many locations have reported -50 degrees F and colder. Exceptionally harsh winters of 1899, 1909, 1912, 1936, 1966, and 1996 brought such temperatures to the area. Two of the most prolonged harsh winter periods occurred in 1911–12 and 1935–36.

From December 25, 1911, to February 14, 1912, the park was consistently in the grip of arctic air. During that time 55 of 57 days produced overnight lows below 0 degrees F. Ten nights were as cold as -40 degrees F, and four nights were -50 degrees F or colder. In the first half of January 1912 climate records show 360 consecutive hours recorded below 0 degrees F. Since

blocks of ice from Namakan Lake were stored in an ice house to provide refrigeration for the hotel built at Kettle Falls in 1910, one could imagine that there was plenty of ice to harvest during that severe winter.

January 5, 1935, to February 22, 1936, brought another prolonged arctic outbreak to the area. During that period, 47 of 48 days brought overnight lows below 0 degrees F. There were 11 mornings of -40 degrees F or colder, bottoming out at -52 degrees F on January 23. With over 24 inches of snow cover, the area endured 240 consecutive hours below 0 degrees F during January, and another 216 consecutive hours below 0 degrees F in February. Once again the lake ice thickness was exceptional that winter. Martin Hovde, the section director for the Weather Bureau wrote "ice of unusual thickness was harvested that winter . . . and provided ice house refrigeration well into the summer months." In fact ice-out on Rainy Lake did not occur until May 13, 1936.

The last major arctic outbreak in Voyageurs National Park occurred over the period from January 19 to February 4, 1996, bringing many overnight lows of -40 to -50 degrees F. Extra thick lake ice was observed on Rainy Lake where ice-out did not occur until May 3, 1996. In fact, there was still floating ice in the open waters of Rainy Lake over the fishing opener on May 11.

Freezing rain is not a common occurrence in Voyageurs National Park, averaging only about three days per year. But December is the month most likely to bring freezing rain. This happens when southerly winds bring warmer air aloft that can override air that is below freezing near the surface. Small droplets precipitating from the base of the warmer clouds fall through the colder air and freeze upon contact with the rocks and vegetation on the landscape of the park. This can make for both beautiful and treacherous conditions. Blizzards and cold waves can occur in December as well. Significant December blizzards have occurred with a frequency of about once every five years. A blizzard over December 12 and 13, 1968, brought everything to a standstill as up to 14 inches of snowfall was blown around by 30 to 40 mph winds into drifts several feet high. Another blizzard from December 29 to 31, 1972, brought over 50 consecutive hours of snow and blowing snow producing drifts that were several fee high. In both 2008 and 2009 blizzard-like conditions occurred during December, the last episode coming at Christmastime from December 24 to 26, 2009, when 30-plus mph winds blew a 15-inch snowfall into huge drifts. December of 1983 will long be remembered as the coldest in modern history averaging well below 0 degrees F for the entire month. The week leading up to Christmas that year produced wind chill index values as cold as -55 degrees F. Engine block or oil pan heaters were commonly used in cars and trucks that December so they would still start and not freeze up.

Extreme measures of December climate include a snowfall of 12 inches on December 13, 1968; an afternoon high of 59 degrees F on December 6, 1939; a frigid low of -51 degrees F on December 28, 1933; and daily precipitation of 1.21 inches on December 5, 1960. The driest December in history brought just 0.07 inches in 1913. The wettest December was in 1918 when 3.20 inches was reported and there were 15 days with significant precipitation. December of 1992 brought a record 43.9 inches of snowfall, with measurable amounts reported on 20 days.

Based on historical averages, January is the snowiest month of the year in Voyageurs National Park, typically ranging from 12 to 16 inches of new snowfall. Average maximum snow depth often occurs in late January There have been snowy winters when snow depth has exceeded 30 inches, making it mandatory to wear snowshoes when moving around outdoors. Such was the case in the winters of 1915–16 (at least 40 inches snow depth), 1916–17 (at least 32 inches snow depth),

1955–56 (at least 36 inches snow depth), 1965–66 (at least 38 inches snow depth), 1971–72 (at least 36 inches snow depth), 1981–82 (at least 38 inches snow depth), and 1995–96 (at least 36 inches snow depth).

Typically the coldest temperature readings of the year occur in mid-January. Many locations in and around the park have reported -50 degrees F and colder. Unusually harsh winters of 1898–99, 1908–09, 1911–12, 1935–36, 1965–66, and 1995–96 brought such temperatures to the area. In January of 1912, 1936, and 1966 some daytime high temperatures remained colder than -15 degrees F all day. January can be dominated by arctic high-pressure systems that bring bright blue skies and light winds. But under such conditions it is often difficult to see because of the amount of reflected light coming off the snow cover. Relatively fresh and powdery snow cover can reflect 80 to 90 percent of the sunlight, almost blinding for most people. As a result sunglasses are commonly worn in the mid winter months to protect the eyes and allow better vision, especially while driving on the ice roads of the lakes.

Extreme measures of January climate include a snowfall of 14.1 inches on January 10, 1975; an afternoon high of 54 degrees F on January 23, 1942; a frigid low of -55 degrees F on January 6, 1909; and daily precipitation of 2.70 inches on January 22, 1982. The driest January in history brought just a trace of precipitation in 2003. The wettest January was in 1975 when 3.03 inches was reported and there were 16 days with significant precipitation. January 1975 also produced 43 inches of snowfall, a record amount.

February tends to be dominated by high pressure with less storminess. On average it is the driest month of the year. Blizzards and heavy snows are fewer in number during this month. As a result, February brings the highest frequency of clear skies to Voyageurs National Park, so there are generally several sunny days. Oftentimes the more powerful sun of February will begin to melt snow that is not protected by shade. This differential melting of the snow cover will leave some odd-looking snow sculptures on the landscape, sometimes taking the form of pillars or stalagmites. Daytime high temperatures more frequently reach above freezing (32 degrees F) during February, but they can still plummet well below 0 degrees F at night. On occasion the daily temperature can range over 50 degrees F. Since February is the least cloudy of the winter months and the low sun angle illuminates the lakes and landscape in such contrasting hues, many photographers attempt to capture vivid images amid the quiet solitude during the daytime. In addition under the clear nighttime skies many will visit and attempt to photograph the northern lights (aurora borealis) when they appear.

Extreme measures of February climate include a snowfall of 12.1 inches on February 27, 1996; an afternoon high of 62 degrees F on February 26, 1958; a frigid low of -48 degrees F on February 8, 1909, and February 1, 1996; and daily precipitation of 1.20 inches on February 1, 1911. The driest February in history brought just a trace of precipitation in 1909. The wettest February was in 1911 when 2.90 inches was reported and there were 7 days with significant precipitation. February of 1955 brought a record 32.3 inches of snowfall, the only time over 30 inches of snowfall came to the park in that month.

An icy point is slowly covered by drifting snow on this cold January afternoon.

From late November through early April, Voyageurs sleeps under a deep mantle of snow. The wind sculpts the snow into interesting drifts along the shoreline of the park's many islands.

On an open lake where ice depths can exceed three feet, the wind sweeps the ice clear, leaving a hard glassy surface.

Spray from crashing waves during a late fall storm created these ice formations on the northwest end of Chase Island on Kabetogama Lake. The cold temperatures and limited sunlight preserve the ice formations all winter.

The afternoon sunshine highlights the snow texture on the shore of Bowman's Island on Kabetogama Lake. In January the sun is so low that this lighting occurs as early as 4:00 p.m.

Winter temperatures in Voyageurs can drop to -40 degrees F or lower and snowfall amounts average 60 to 75 inches. In extreme winters, snowfall accumulations can reach 120 inches.

The water level on the main lakes drops during the winter. The water level is regulated by dams at Kettle Falls, Squirrel Falls, and International Falls to ensure a continuous flow of water through the hydro plant in International Falls. The level changes create pressure ridges across the lakes. Cracks in the ice provide openings for otter to get in and out of the water in the winter. The cracks also allow water to seep up over the ice, forming patches of slush under the snow that don't refreeze and can be dangerous for snowmobilers and skiers.

Ice stalagmites formed by the spray from late fall storms contrast with orange lichen on an island on the east end of Namakan Lake.

A deer browses for food on a winter afternoon. Deep snow, wolves, and lack of food make winter life for deer precarious at best.

BELOW: A deer killed by timber wolves lies partially eaten on Kabetogama Lake. Deer make up approximately 80 percent of a timber wolf's diet.

Ravens keep a keen eye out for carrion and often compete with wolves for the remnants of a dead deer.

BELOW: The pattern in snow reveals where a ruffed grouse spent the night. The birds dive under the snow for warmth and to evade predators in the winter.

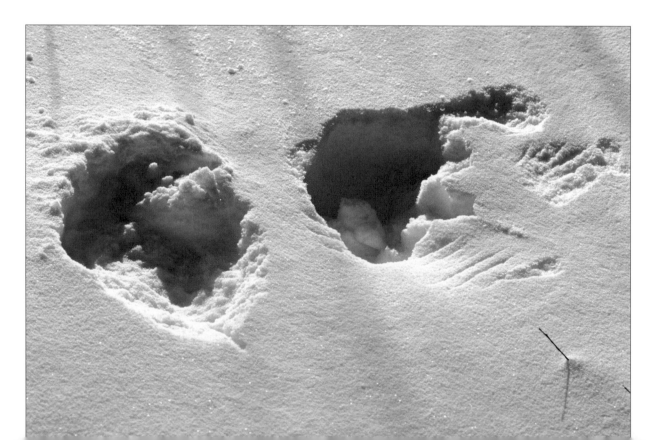

Ermine tracks traverse a bay on Namakan Lake. One of the pleasures in traveling the park in the winter is finding tracks that reveal the presence of animals one seldom sees.

The scarlet plumage of a male pine grosbeak brightens a dark winter day. Voyageurs is in the heart of the wintering range of these far northern birds.

BELOW: Advection or wind frost decorates a jack pine on the shores of Rainy Lake. Advection frost is a form of hoar frost that occurs when humid weather is followed by a cold wind. Note the tiny ice spikes that form on the downwind side of the needles.

The rugged shoreline stands defiantly against the heavy forces that nature hurls against the eastern end of Namakan Lake. Heavy northwest winds, both winter and summer, take their toll on the struggling vegetation.

Northwest winds create snowdrifts that mimic summertime whitecaps on the east end of Namakan Lake. Northwest winds prevail from November through March.

A parade of roots is all that remains of trees that were flooded out when the Kettle Falls dam raised the lake level in the Namakan basin.

The stable winter air allows clouds to form long streaks above Williams Island on Namakan Lake.

A stunted white pine holds a lonely winter vigil against the backdrop of a mackerel sky. This high cloud pattern is a common occurrence in winter.

OUTLOOK

RESIDING ALONG THE BORDER OF TWO COUNTRIES, centered in the middle of the North American continent, and at a midpoint between the Pacific and Atlantic Oceans, Voyageurs National Park understandably has one of the most variable climates in the national park system. Weather events and episodes can be beautiful, frightful, dramatic, and traumatic all in the same day. Further, the park is large enough that while severe weather might be occurring over one lake, another lake might be perfectly tranquil. Within this highly variable weather behavior it is difficult to sort out significant trends, but there are indeed signs that the climate of Voyageurs National Park is changing and is already having consequences on its ecosystems.

During the three most recent decades (1980s, 1990s, and the first decade of the new millennium) Minnesota's climate has shown four very significant trends, all of which have had many observed impacts. These four trends are

statistically detectable in the data of most Minnesota climate stations, including those nearby and within Voyageurs National Park. These trends include (1) warm winters—both in persistence and amplitude of the positive temperature departures, (2) higher minimum temperatures, (3) increased frequency in episodes of high summer dew points—a measure of atmospheric water vapor, and (4) greater variation in annual precipitation—most profoundly in thunderstorm rainfall. For those who

manage Voyageurs National Park, as well as those who enjoy visiting, these trends may be worth noting in detail.

The statistical signal for warm winters appears in a number of ways. By examining the heating season (November through March) mean monthly temperature value patterns can be discerned. Some of these climate records go back to 1897; so long term trends can be evaluated as well. Twelve of the 24 warmest heating

Altostratus clouds loom over Little Chase Island, Kabetogama Lake on a warm spring evening.

seasons for Voyageurs National Park have come in the past two decades (since 1990). Further, a ranking of the core months of winter (December through February) shows that seven of the warmest thirteen have occurred since 1994. The distribution of daily temperatures for Voyageurs National Park shows that a vast majority of winter days in the past decade have provided temperatures that are above historical averages. January 2006 was the warmest in the historical record and on 15 days during that month the daily temperature was 20 degrees F or greater above normal, peaking at 44 degrees F on the 27th. The upward trend in mean annual temperature for Voyageurs National Park is heavily weighted to the significant contributions from the winter months and the heating season overall (November through March).

Observable impacts of this warm winter trend are evident to most nearby residents and visitors. There has been a detectable change in the onset of ice cover on lakes, the thickness of the ice, and the loss of ice cover in the spring. In other words, the ice fishing season has been shorter overall in recent decades. Examining the Minnesota Department of Natural Resources data for ice-out dates on Rainy Lake shows a trend for earlier spring dates. The historical average ice-out date for Rainy Lake since 1930 has been May 5. Since 1990 the average ice-out date has been May 1 and the two earliest ice-out dates in the historical record were April 13, 1998, and April 10, 2010. Research from the University of Minnesota indicates that as climate continues to change the overall ice cover season on the lakes may be reduced by as much as two to three weeks.

A modest upward trend in mean annual temperature can be found in the climate station records for those locations in and nearby Voyageurs National Park. This upward trend in mean annual temperatures appears to range from 0.3 to 0.4 degrees F per decade. As mentioned above this trend is heavily weighted to the warm winter months, but it is also weighted towards a greater positive change in overnight winter minimum temperature than winter daytime maximum temperature. Examining the temperature records at International Falls shows that for the period from December through March the average daytime maximum temperature values have warmed by 3 degrees F while the average overnight minimum temperature values have warmed by 4 degrees F. Temperature changes in the other months of the year have been relatively modest and in some cases slightly negative (cooler). The frequency of overnight lows of -35 degrees F or colder has diminished from three to four times per year to just one or two times per year in recent decades.

So what are the impacts of relatively higher minimum temperatures? At this point it is unclear what significant consequences from this trend may be occurring specifically in Voyageurs National Park. Perhaps new biological patterns will emerge in future years as a result of higher minimum temperatures. Many soil microbes, plant pathogens, and insects survive and even thrive better when minimum temperatures are higher. Wildlife biologists also tell us that animal migration, hibernation, and winter foraging behaviors can be altered by warmer nighttime temperatures.

One result from the higher minimum temperatures in the spring (April and May) and the autumn (September and October) has been a change in the freeze-free growing season. Studies have shown that the average last spring freeze date (32 degrees F or colder) for most Minnesota climate stations has come earlier on the calendar in recent decades. The average first fall freeze date has been displaced to later in the season as well, but fewer days than the change in spring freeze dates. The end result is an extension of the freeze-free growing season across Voyageurs National Park, adding an extra seven to ten days.

The third trend involves a climate attribute called dew point. Dew point is a measure of atmospheric water vapor. The higher the dew point the greater the abundance of water vapor in the air. Because water vapor is the primary greenhouse gas in the Earth's atmosphere, it has a significant effect on air temperature, particularly during the night when the air stratifies into stable layers and there is less mixing. For much of Minnesota it appears that the frequency of tropical-like dew points (70 degrees F or higher) is increasing. At Voyageurs National Park the frequency of 70 degrees F dew points or higher in the past decade (since 2000) is nearly twice the average for the previous five decades. From 1950 to 1999 there was an average of four to five days per year when the dew point reached 70 degrees F or higher. Since 2000, the annual average has been eight to nine days per year.

Dew points of 75 degrees F are common in the Persian Gulf, low latitude coastal climates, and tropical rainforest climates, but they are rare in Minnesota's northern landscape. From 1948 to 1976 the International Falls climate record shows a frequency for dew points of 75 degrees F or higher that represents less than one time per year (20 times in 29 years). But since 1977, the climate record shows a frequency of approximately four times per year (90 times in 23 years). Further, some summers have included spells of extremely oppressive weather with dew points above 75 degrees F and temperatures

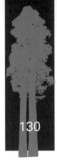

in the 90s F, resulting in Heat Index Values (a measure of how hot the combination of temperature and dew point make the air feel) that ranged from 100 degrees F to 108 degrees F. These spells occurred in the summers of 1977, 1996, 2001, and 2005. Heat Index Values of 105 degrees F or higher are detrimental to human health and can lead to heat exhaustion or heat stroke. As such the National Weather Service is obligated to issue a heat advisory when the Heat Index is expected to reach 100 degrees F for three hours or more. Since 1988 every heat advisory issued has been associated with spells of high dew points combined with air temperatures of 90 degrees F or higher.

Higher nighttime minimum temperatures are associated with high dew points, particularly the number of times the temperature never drops below 70 degrees F making it quite uncomfortable and difficult to sleep for those camping or staying in cabins without air conditioning. Temperature records from in and around Voyageurs National Park show only eight nights when the overnight low remained at 70 degrees F or higher from 1948 to 1976, while fifteen such nights have occurred from 1977 to 2009. In the summers of 2001 and 2002 there were some nights when the overnight low never fell below 75 degrees F, something that had not occurred since the heat wave and Dust Bowl era of the 1930s.

Observable impacts of the trend toward high dew point spells in the summer months are wide ranging. One of the impacts has been increased water temperatures in some of Voyageurs' lakes, especially the shallower and more nutrient rich Lake Kabetogama. Higher water temperatures have implications for water quality, particularly plant growth and fish population changes. Elevated water temperatures in the July and early August period can lead to more productive algal blooms in some bays and shallows, resulting in less clarity in the water, which normally ranges from 8 to 14 feet. From annual water quality measurements taken over many years, Voyageurs' lakes have nutrient and clarity values that make them oligotrophic (clear, low productivity) to mesotrophic (intermediate productivity and clarity) in most summers. However, in warm summers with higher dew points, Kabetogama and Black Bay have at times been classified as eutrophic (high productivity, and diminished clarity) when clarity values have fallen below 5 feet. The Minnesota Department of Natural Resources and Pollution Control Agency continue to monitor the water quality with Voyageurs National Park to detect patterns of behavior that may be related to climate change.

The last, but certainly no less significant, recent climate trend in Voyageurs National Park is more variability in precipitation, with an associated change in the character of storms. Climate stations associated with Voyageurs

National Park show unusually wet and dry years to be more common in occurrence over recent decades. Certainly 1977, 1985, 1999, 2004, and 2005 are among the wettest years in history, ranging from six to ten inches above average annual precipitation. Kabetogama reported 40.56 inches in 1985, a record annual amount for Voyageurs. Conversely, 1976, 1990, 2003, and 2006 are among the driest years in history, ranging from five to seven inches below the average annual precipitation. Over recent decades, even within-year variability of precipitation has been remarkable as well. In 2004 Kettle Falls reported 24.85 inches of precipitation, while Kabetogama reported 36.39 inches, over 11.5 inches more across a distance of just 21 miles. With this high degree of spatial variability it is not inconceivable that in the future one part of the park may be in drought, while another area of the park may see a precipitation surplus.

What's behind this amplified variability in precipitation? One reason for the higher variability is that a larger fraction of the annual precipitation is coming in the form of thunderstorm rainfall. The evidence for this is found in the frequency of daily rainfall amounts of one inch or greater. This level of precipitation is almost always associated with convective thunderstorms and primarily occurs during the period from May to September across Voyageurs National Park. Some climate stations in and nearby the park show an increased frequency of 1-inch rainfall amounts over the past two decades. Kettle Falls, Kabetogama, Littlefork, and Big Falls climate records confirm this trend. The high degree of year to year variability in precipitation, as well as a periodic large spatial variability in precipitation across the park present problems for the management of water levels in the lakes, something that is controlled at the Kettle Falls dam.

In summary, the climate is changing in Voyageurs National Park. These changes are a manifestation of higher water vapor in the atmosphere over Minnesota. The combined attributes of being a source of latent heat and a greenhouse gas imply that increased water vapor presence favors warmer winters, higher minimum temperatures and dew points, and a greater contribution from thunderstorm rainfall. It appears that these climate trends are not about to reverse themselves soon and are not occurring in isolation over Voyageurs National Park. On the contrary these trends are in evidence across most of the landscapes in the Western Great Lakes Region. Greater attention to these climate trends and their implications may be necessary to adapt and manage Voyageurs National Park to ensure that its pristine waters and beautiful landscapes are preserved for future generations.

PHOTOGRAPHER'S NOTES

OVER THE YEARS I HAVE LEARNED A GREAT DEAL about photographing in the outdoors. I hope these suggestions will help you with your photography. The photos in this book span a period of about 40 years and a range of cameras from 35mm to 4 x 5 to digital. Although I do not give specific exposure and film detail about each shot, I would like to offer some general information about equipment and working techniques that might be helpful to other photographers.

I began photographing with a Pentax 35mm camera equipped with a normal lens and some supplementary close-up lenses in the mid 1960s. Later, I moved to a Leica M2 camera and a Nikon F camera. Subsequently, I switched to a Leicaflex and also used a Mamyia RB 67 medium format camera. I was dissatisfied by the lack of depth-of-field with the RB 67, so I traded it for a 4 x 5 field camera so I could use the swings and tilts to enhance depth-of-field.

In 2004 I began using a digital camera and soon retired my film cameras! The ease of working in the digital format, the unlimited number of images that can be made at no cost, the ability to instantly review your work, plus the control the photographer can have over the final image give digital an overwhelming advantage.

Currently, I use Canon 5D and Canon 7D cameras. The 5D is a full format camera (e.g., the image sensor is the same size as a 35mm film frame). The 7D has an APS sized sensor which is about 2/3 the size of a 35mm film frame. I use the 5D with shorter lenses to take advantage of the full frame camera's ability to get the full wide angle effect of shorter focal length lenses. I use the 7D with long lenses because the smaller sensor has a magnifying effect on the image of 1.6X. Thus, a 400mm lens produces the same image magnification that a 640mm lens would produce on a full frame camera like the 5D. This is a distinct advantage when photographing wildlife. I usually have a 100–400mm zoom lens mounted on the 7D and a 24–105mm zoom lens mounted on the 5D. That keeps lens changing to a minimum while on water. Other lenses I use frequently are a 24mm tilt shift lens that gives some of the same control over perspective and depth-of-field as a view camera, and also a 70–200mm zoom For close-ups, I use a supplementary close-up lens on the 70–200mm lens and a 50mm extension tube with other lenses. Although this may not be the ideal kit for close-up photography, it works and I don't have to carry another lens.

For shooting on water, I have used a variety of boats and canoes. My favorite is a 16-foot aluminum fishing boat equipped with a 25-horsepower motor and an electric trolling motor. Wildlife are used to seeing boats in the park and the boat provides a much steadier shooting platform than a canoe. The electric motor moves quietly and does not spook wildlife. When I see something that I want to photograph, I maneuver the boat into position, shut the motor off, and shoot until the boat drifts out of position. Even on calm days, a boat never holds its position very long. Working from a boat means handholding the camera. Lenses with built-in image stabilizers help get sharp images when shooting from a moving boat. With a 100–400mm lens, I often use a shoulder stock to steady the camera as well.

Using a small boat also allows me to land and go ashore nearly anywhere. Once on shore I usually use a tripod to steady the camera. Using a tripod slows you down and allows you to study the image in the viewfinder and to make subtle adjustments to the composition.

I prefer a carbon fiber tripod and a ball head with a mounting plate for the camera. This allows fast set up and rapid adjustment of the camera position. The tripod has adjustable leg stops to vary the spread of the legs for the uneven ground. The leg stops also allow the tripod to be positioned at ground level, a necessary requirement for wildflower photography.

Winter presents its own set of challenges for the photographer. Film freezes, batteries die, and fingers get numb! The whiteness of the snow causes underexposure by the camera's light meter. To compensate for the whiteness, I increase the exposure by one stop over the camera's recommend setting. By so doing, I get white landscapes instead of muddy grey ones. Often, in difficult lighting conditions, I bracket my exposures—shooting several different exposures, knowing that I will toss out the bad exposures later.

Digital cameras work well in reasonably cold weather; I have shot up to two hours in -10 degree F temperatures without the camera shutting down. Not having to change film every 36 shots with cold stiff fingers and not having to worry about cold film breaking in the camera are distinct advantages of digital cameras!

In deep snow, setting up a tripod can be a problem because you can't get the tripod legs down to a solid footing. Instead, I use a monopod because it can be shoved down through the snow to stabilize the camera.

Avoid breathing or blowing on a cold camera because it will fog the viewfinder or lens. I carry a shaving brush to clean snow off the camera and lens.

After shooting in cold weather, don't take a cold camera or lens directly into a warm room. If you do, condensation will form on the equipment and can damage the electronic components. I put my cold equipment into a zip lock plastic bag before taking it inside. The bag keeps the condensation off the equipment until it warms up.

I photograph both RAW and JPEG files, process the images in Photoshop on a Mac computer, and print on an Epson inkjet printer. I save the images on a stand-alone hard drive and also on DVDs.

I hope that these technical comments and tips are helpful to you.

HAVE FUN!

Cold temperatures and blinding white snow make winter photography a bit tricky. Carry extra batteries and compensate for snow brightness by increasing your exposure by one stop above your camera's recommended exposure.

GLOSSARY

ADVECTION: The transport of an atmospheric property (typically temperature or moisture) solely by the mass motion (velocity) of the atmosphere. Most often in common weather usage advection refers only to the horizontal large-scale motions of the atmosphere, while convection describes the predominantly vertical, locally induced motions.

ALGAL BLOOM: The rapid excessive growth of algae that is generally caused by high nutrient levels and favorable temperature conditions in a water body. This can result in deoxygenation of the water mass when the algae die.

CARRION: usually a dead and decaying animal that may be eaten by other animals such as vultures.

CONIFEROUS: refers to types of trees that are cone-bearing and exhibit evergreen leaves (needles).

CONVECTION THUNDERSTORMS: a vertical cloud mass, usually in the form of cumulonimbus (sometimes anvil shaped) produced by the vertical transport of heat and moisture in the atmosphere. These storms produce lightning, thunder, gusty winds, and intense rainfall at various spatial and temporal scales.

CUMULUS CLOUDS: one of the principal cloud types that shows distinct vertical dimensions and appears as mounds, domes, or towers. They may be isolated or in clusters.

DECIDUOUS: plants, shrubs, and trees that shed their leaves at the end of each growing season and go through a dormant period without leaves.

DERECHO: a large-scale, highly organized convection thunderstorm that produces damaging straight-line winds, often cutting a wide swath of damage across the landscape.

DEWPOINT: commonly a lower air temperature at which a given amount of air, unchanged in pressure and moisture content, will reach saturation and therefore the water vapor will condense into droplets (dew).

DIURNAL: an adjective that refers to a daily cycle or range in a weather or climate attribute (temperature for example).

DUST BOWL: the name given to the central region of the USA that was stricken with severe drought and dust storms during the 1930s.

ECOSYSTEM: living organisms, their environment, and how they interact.

EQUINOX: most commonly the time of year when the midday sun passes directly over the equator, on or about March 21 on its way into the northern hemisphere, and on or about September 21, when it is on its way into the southern hemisphere.

EUTROPHIC: often used to describe the condition of a lake or pond which is rich in mineral and organic nutrients that promote plant life, especially algae. This can reduce the dissolved oxygen content and cause problems for a diversity of aquatic organisms.

GREENHOUSE GASES: atmospheric gases, such as water vapor, carbon dioxide, methane, nitrous oxide, and ozone, among others that are transparent to incoming shorter wavelengths of solar radiation (sun's energy), but are efficient at trapping and storing the infrared radiation emitted from the Earth and atmosphere. The trapped infrared radiation (longer wavelength) controls the Earth's surface temperature.

HEAT INDEX: this term is derived from the research of Robert G. Steadman who showed the stressful effects on the human body caused by combinations of high temperature and high dewpoint (or humidity). When the air temperature is above 80 degrees F and relative humidity is above 40 percent a Heat Index value can be calculated from the combination of temperature and dewpoint (humidity conditions) to assess what the equivalent temperature feels like on the human body if the humidity was below 40 percent. For example an air temperature of 90 degrees F with a relative humidity of 60 percent produces a Heat Index of 100 degrees F. The National Weather Service uses Heat Index values as criteria to issue heat watches and advisories to the public as values above 105 degrees F are detrimental to human health.

HERBACEOUS: refers to a type of plant that is not woody, but leafy and herblike.

ICE-OUT: in Minnesota this most often refers to a spring date when 90 percent of the lake ice is gone or enough ice is gone that boating navigation is possible.

LEE SIDE: in meteorology this refers to the downwind (rear) side of an object relative to the atmospheric flow (wind). For example in the case of a wind from the west passing over a mountain barrier, the lee side would be on the east slope of the mountain.

LICHEN: any of various small complex plants composed of a fungus and an alga growing in a symbiotic (cooperative) relationship on a solid surface like a tree or rock.

MEAN ANNUAL TEMPERATURE: most climatologists calculate this value of temperature by averaging the 12 monthly values of mean temperature for the most recent three complete decades. Thus 1971–2000 represents the 30-year mean period for current calculations.

MESOTROPHIC: often used to describe the condition of a lake or pond as moderate in nutrient content such as nitrogen or phosphorus. This tends to be more common among shallow lakes instead of deeper ones.

OLIGOTROPHIC: a term used to describe a lake or pond that is very low in nutrient content. These waters have a low level of biological productivity and are generally relatively clear.

PREVAILING WIND: the wind direction that is most frequently observed during a specified period of time. This is noted as the direction from which the wind blows (e.g., northwest denoted as NW).

RADIANT ENERGY: energy that is propagated in wave form through the atmosphere. It may be shorter wavelengths (such as the visible spectrum) or longer wavelengths (infrared spectrum). Radiant energy can come directly from the sun, the sky, clouds, or the Earth itself.

RAFTED ICE: also called telescoped ice, this is ice that is adrift in pieces that override each other and sometimes forms stacks. It can especially be common during late winter and early spring when ice is breaking up.

SOLSTICE: either of two dates on the annual calendar when the vertical midday sun is displaced farthest, north or south, from the Earth's equator. This is over the Tropic of Cancer (23.5 degrees north latitude) in the northern hemisphere summer about June 21, and over the Tropic of Capricorn (23.5 degrees south latitude) in the southern hemisphere about December 21.

SPATIAL: of or pertaining to the areal distribution of any given quantity, for example temperature or rainfall, distributed across a landscape.

STALAGMITES: most commonly a cone-shaped deposit of calcium carbonate found on the floor of a cave underneath a dripping source of calcareous water.

STRATOFORM: an adjective used to describe low to mid-level cloud types that form a somewhat continuous layer like a blanket. As opposed to vertically developed clouds that are called cumuloform.

STRATUS CLOUDS: a horizontally layered cloud form (like a blanket) with a uniform base. Light precipitation or drizzle may be associated with these clouds.

SUN DOGS: also referred to as parhelion this takes the form of a colored spot (halo) at or near the same elevation as the sun and often on either or both sides. It is caused by refraction of light through suspended ice particles in the atmosphere.

TEMPERATURE INVERSION: a layer of air where temperature increases with elevation. This type of air mass does not allow for vertical mixing of air and can lead to fog, haze, or poor air quality.

TEMPORAL: of or pertaining to the distribution of some quantity over time, for example the daily or monthly variation in temperature.

UNDERSTORY PLANTS: most commonly the shade-tolerant vegetation that grows underneath a forest canopy. These may be grasses, forbs, or shrubs.

WIND SHEAR (ALOFT): a change in wind direction and/or speed with height. Sometimes this change with height can be significant so that a low layer of clouds is seen to move in an opposite direction from a higher layer of clouds.

WIND-CHILL INDEX: a measure of the cooling effect caused by the combination of air temperature and wind. The index relates to the rate of heat loss from exposed human skin. When index values are colder than -20 degrees F there is a threat of frostbite to exposed skin

ACKNOWLEDGMENTS

MANY PEOPLE HAVE COME TOGETHER to make this book possible. I thank my co-author, Mark Seeley, for his unwavering enthusiasm. Mark's rich knowledge about climate and weather patterns in Voyageurs National Park adds a depth to this book that makes it more than a coffee-table book of nature photos. He also worked tirelessly to garner publishing and financial support for our work.

Afton Press publisher, Patricia Condon McDonald, provided valuable insight and guidance throughout the writing, design, and publication of this book. She also played a vital role in securing funding. Mary Sue Oleson, our talented designer, did a magnificent job interpreting my vision for the book; and Beth Williams, copyeditor and production assistant, who kept everything on track.

Voyageurs National Park superintendent, Michael Ward, and his staff were willing to review the content of the book and offered substantive information on the cultural and natural history of the area.

I would also like to thank Gene Merriam, President of the Freshwater Society, for writing the foreword and Faye Sleeper, Co-Director of the University of Minnesota Water Resources Center, for her endorsement.

My wife, Lynne, was always supportive and willing to give me the time necessary to make the images presented in this work. She also served as an untiring sounding board for ideas. My parents, Floyd and Ruth Breneman, introduced me to the outdoors and to the country of Voyaguers National Park at an early age. They instilled in me a deep respect for nature and supported my love of photography.

I also acknowledge David Trappe, my high school history teacher, who started a photo club in our school that nurtured my serious interest in photography.

Finally, I want to thank the many sponsors whose generosity made the publication of this book possible.

OPPOSITE: The late evening sunset silhouettes a dead white pine on one of the hundreds of rocky points in Voyageurs National Park. Weather and disease take a toll on these magnificent trees each year.

THIS BOOK WAS DESIGNED
WITH CARE BY

Mary Susan Oleson
Nashville, Tennessee